MW00718369

$14.75

PowerKids Readers:
Nature Books™

Rainbows

Jacqueline Dwyer

The Rosen Publishing Group's
PowerKids Press™
New York

1

For Gareth

Published in 2001 by The Rosen Publishing Group, Inc.
29 East 21st Street, New York, NY 10010

First Edition

Book Design: Michael de Guzman
Layout: Felicity Erwin

Photo Credits: p. 1 © International Stock/Orion; p. 5 © F.P.G./Ron Chapple; p. 7 © F.P.G./Haroldo De Faria Castro Cast; p. 9 © International Stock/Chad Ehlers; p. 11 © International Stock/Warren Faidley; p. 13 © F.P.G./Jeff Divine; p. 15 © F.P.G./Willard Clay; p. 17 © F.P.G./Charles Benes; p. 19 © Index Stock Photography, Inc.; p. 21 © International Stock/Bill Tucker; p. 22 (rainstorm) © F.P.G./VCG, (sunshine) © F.P.G./Frank Cezus.

Dwyer, Jackie, 1970-
    Rainbows / by Jacqueline Dwyer.
        p. cm.— (PowerKids readers. Nature books)
    Includes bibliographical references and index.
    Summary: Describes what rainbows are made of, what colors they contain, and how readers can make rainbows of their own.
    ISBN 0-8239-5676-8 (lib. bdg.)
    1. Rainbow—Juvenile literature. [1. Rainbow.] I. Title. II. Series.

QC976.R2 D79 2000
551.56'7—dc21   99-042984

Manufactured in the United States of America

2

Contents

Rainbows are made from drops of water and sunshine.

5

Rainbows are colored light. The light comes from the sun's rays that shine through drops of water.

Some rainbows appear in the sky after a big rainstorm.

There are many colors in a rainbow. The colors in a rainbow are red, orange, yellow, green, blue, and purple.

red

orange

yellow

green

blue

purple

Rainbows make a special shape. This shape is called an arc.

13

Some rainbows are big. Other rainbows are small. This rainbow is bigger than a house!

Sometimes you can see
two rainbows in the sky.

17

On a sunny day,
you can make your
own rainbows.

rainbow

19

It is fun to paint rainbows.
You can use red, orange,
yellow, green, blue, and
purple to make your
rainbow.

21

Words to Know

ARC

PAINT

RAINBOW

RAINSTORM

SKY

SUNSHINE

Here are more books to read about
rainbows:
All the Colors of the Rainbow
(Rookie Read-About Series)
by Allan Fowler
Children's Press

Raindrops and Rainbows
by Rose Wyler
Julian Messner

To learn more about rainbows, check out
these Web sites:
http://www.brainpop.com/rainbow
http://www.deltatech.com/rainbowx.html
http://www.zianet.com/rainbow/frmake.htm

Index

Word Count: 121

Note to Librarians, Teachers, and Parents

PowerKids Readers (Nature Books) are specially designed to help emergent and beginning readers build their skills in reading for information. Simple vocabulary and concepts are paired with photographs of real kids in real-life situations or stunning, detailed images from the natural world around them. Readers will respond to written language by linking meaning with their own everyday experiences and observations. Sentences are short and simple, employing a basic vocabulary of sight words, as well as new words that describe objects or processes that take place in the natural world. Large type, clean design, and photographs corresponding directly to the text all help children to decipher meaning. Features such as a contents page, picture glossary, and index help children get the most out of PowerKids Readers. They also introduce children to the basic elements of a book, which they will encounter in their future reading experiences. Lists of related books and Web sites encourage kids to explore other sources and to continue the process of learning.

24